P9-CDB-679

OTHER BOOKS BY JOHN CANADAY

Metropolitan Seminars in Art
Mainstreams of Modern Art
Keys to Art (*with Katherine H. Canaday*)
Embattled Critic
Culture Gulch
Lives of the Painters

THE ARTFUL AVOCADO

John Canaday

THE ARTFUL AVOCADO

1973

DOUBLEDAY & COMPANY, INC.

GARDEN CITY, NEW YORK

Designed by Earl Tidwell

Library of Congress Catalog Card Number 72-89820
ISBN: 0-385-00156-8

Printed in the United States of America
First Edition

Dedication and Hello

This book is dedicated to all those people who yearn to grow an avocado bush indoors but think they can't. They just look at you blankly when you tell them how to do it and say, "Yes, I've tried all that," but have never managed to get the hang of successfully midwifing a sprout from one of those fascinating pits. These people feel like outcasts. They have been reduced to total defeatism. They think they are poison to avocados. It's a variation of "I love avocado, but avocado doesn't love me."

I know how they feel, because I'm that way with fuchsias. Bring me a fuchsia to care for over the weekend and you will find its blossoms lying all over the floor Monday morning.

8

And I love fuchsias. On the other hand, I'd be just as happy if I never saw another poinsettia, but give me one for Christmas, as people are always doing, and it will still be blooming vigorously at Easter. People who think that plants like to be talked to, a widespread belief, haven't noticed that they don't care what you say. Right now on my window sill I have a poinsettia going on its sixth month in blatant health and full flower in spite of the fact that every morning when I water it I say, "Get lost." I would try "Drop dead," but I don't feel that way about plants. Any plant.

It just doesn't stand to reason that some people are poison to avocados any more than it stands to reason, all evidence to the contrary, that I am poison to fuchsias. So, pending somebody telling me how to grow fuchsias, this is me telling other people how to grow avocados. I'm lucky that way, having grown a winner that was the subject of an article in the New York *Times* called "My Wife, My Avocado, and Me."

Response to that article accounts for this book. As art critic of the *Times*, I have a column every Sunday and am accustomed to finding letters in my mail box saying "I hate you," but the avocado article brought a stack of letters from people of all kinds—people who can't even get an avocado pit to germinate and people who claim to have avocado bushes bigger than mine (which is ridiculous)—all of them calling me "Brother."

Apparently the fraternity doesn't object to

a little horticultural pragmatism, so let's make it clear that that's what I practice. My definition of "pragmatic" is, "try it and if it works, that's fine, keep it up and throw away the rule book." The dictionary backs me up in more stilted terms, defining pragmatism as "a system of or tendency in philosophy [for which read 'indoor avocado culture'] which tests the validity of all concepts by their practical results." The practical result in this case is a big avocado bush.

The dictionary also says that "pragmatic" can mean "meddlesome, officious, conceited," etc. I reject that, since this manual on how to germinate the pit and educate the plant is offered modestly by a non-botanist and experimental indoor gardener. The importance attached to avocado culture at this amateur level surprised me. It is a veritable passion in metropolitan areas. In all those letters in response to "My Wife, My Avocado, and Me," not a single person asked for marital advice or offered any. All they could talk about was avocados. With thanks to the New York *Times* for reprint permission, "My Wife, My Avocado, and Me" opens this book to fill you in on the emotional problems that may become involved in what you begin as a pastime.

Contents

THE ARTFUL AVOCADO

My Wife, My Avocado and Me

At about the time we moved into our present apartment, which was twelve years ago, I got home one day in time to rescue an avocado pit from the garbage following my wife's concoction of a bowl of guacamole. She is one of the best guacamole makers outside Mexico but has trouble with a husband who can't stand to see an avocado pit go unplanted. Over the years many avocados have come and gone in our family, but the plant from the pit I rescued on this occasion has endured and grown to such an extent that we are now a *menage à trois*—my wife, my avocado, and me. And although my wife in moments of stress says, "One of us goes—it or me," she backs down when I point out that while she can still get through the door, the avocado can't. It rises six feet above its eighteen-inch pot, stretches out sixty-eight inches at the widest part of its irregular mass, and is supported by a trunk five and three-quarters inches in circumference.

It is also, so far as I know, the only Weeping Avocado in the world. Early in its life, when it had stretched up close to the ceiling while pushing out some irrational weedy-looking branches here and there in typical avocado fashion, I topped it just above the juncture of two of these branches and bent them along with other branches downward, securing them in their new positions with loops of string. The avocado has never tried to continue its upward growth with new branches at the cutoff point, but has kept pushing out new branches from these downward-bent branches that, in turn controlled, now give the plant a thick bushy mass. It is an oddly unconventional mass (avocados are individualists), most of the growth having taken place on one side, but you can't control everything, can you?

At the moment, we are on the verge of an emergency. I have said the pot is an eighteen-inch one, and it is inadequate. The plant's roots are now growing upward through the earth for lack of any other space, but I am unable to find a larger clay pot (I don't like the wooden bucket type) and in any case am not sure I could manage a transferral from the present one. In the meanwhile the plant lives on rations of fish emulsion fertilizer issued every second Sunday morning, which it adores, and drinks three to five quarts of water a day, depending on the weather. It gets a lot of sun for a house plant in the city, occupying as it does a large unobstructed east window.

The building's wall drops down to the East River, and I am told that reflections off the water—which we can see playing over our ceilings—add to the plant's usable light. About this I don't know. I turn the pot at 180 degrees from time to time, which also involves changing the position of the piano to adjust to the shape of the foliage, the piano and the avocado sharing the same area as they do. To make things easier on my sacroiliac and less irritating to my wife's psyche (it's her piano), I tried for a while to compensate on the shady side of the plant by giving it artificial light. Nothing happened.

Either avocados are immune to disease or I have been lucky with this one. The only trouble has been with some white flies brought in with a plant I kept for a so-called friend over a holiday. A neighborhood plant store identified the pests for me from description and sold me a spray and sprayer, but I discovered that when sprayed from the recommended distance the varmints merely flew in a white cloud to a new location, while a closer spray killed the leaves along with the flies. Also, poisonous sprays used indoors make me feel uneasy. So, discarding this apparatus, I developed an ambush technique by which I could destroy numbers of the insects at a time by pinching them on the leaf between thumb and forefinger. (White flies are rather dry little creatures, not repulsive and squashy like most plant bugs.) Too hard a pinch bruised the leaf, but I learned just the right pressure and by pinching in spare moments over several weeks got rid of the last of the intruders.

Until we moved to New York my wife and I lived in houses with gardens, sometimes with countryside nearby for walks. Sometimes now on nights when the city has been too much with me, I get a pillow and lie on my back on the floor looking up into the avocado leaves. That way, you can almost imagine you are in some delightful verdant spot. Which is why city people grow plants in captivity.

You can almost imagine you are in some delightful verdant spot somewhere

Germination, Plain and Fancy

So much for the emotional problems and satisfactions you are likely to encounter. Now to get down to business.

The first thing to know about germinating an avocado pit is which end is up. I have known people brilliant in their own fields but so devoid of rudimentary instincts about plants that they couldn't deduce that the flattish end with a sort of navel centered in it is where the roots will come out, growing downward, while the point opens to release the shoot. I say "point," but some pits are almost globular. The only thing you have to know is which end is down. Need we explain that the other end is up? An avocado can take a lot, but I have never known one to grow planted head over heels. (Perhaps it has happened. Avocados are tough and strong-willed.)

Germinating in Water

The most frequently recommended method of germination is to stick toothpicks in the sides of the pit so that they can support the poor thing on the rim of a glass of water in which its bottom is submerged for about half an inch. This rigamarole works and I am about to describe it, although I long ago abandoned it in favor of planting directly in soil—a method I'll describe later. However, the water-glass method gives you the fun of watching the roots grow, and since half the

Half an avacado, properly opened, ready for removal of pit

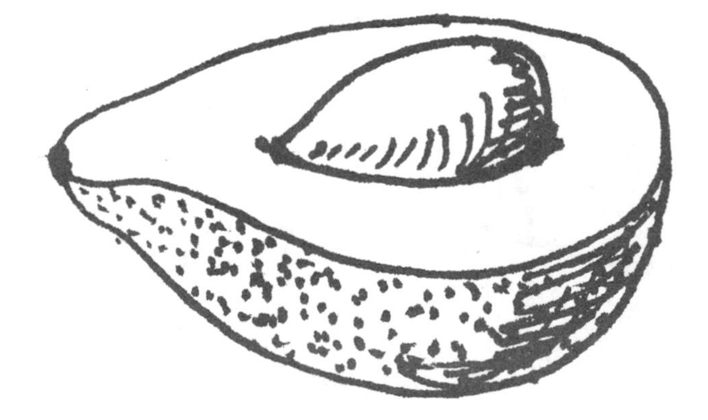

Like a miniature botanical spacecraft

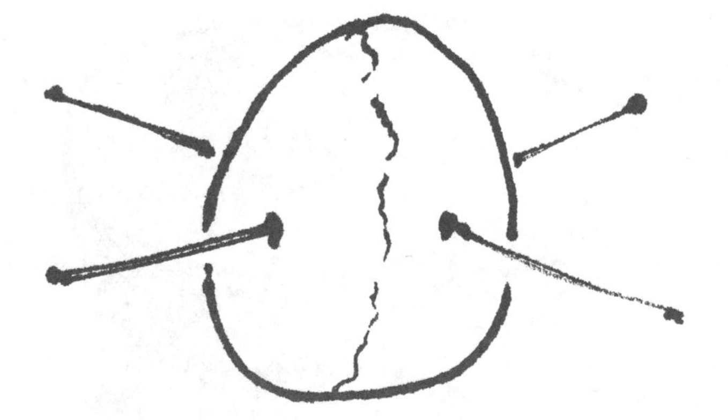

fun of avocado culture is watching things happen, this is an advantage.

So slice your avocado open, extract the pit, and give it a pleasant introduction to this world with a nice warm bath. This is an aesthetic measure rather than a practical one; washing and drying gets off bits of avocado meat that would rot. Then stick four (more if you wish) good strong toothpicks into it as per the illustration here. Although it seems cruel, you could use large needles, which would go in easier. You're not about to wound the pit by going in half an inch or so. The important things happen at its center.

You may take off some or all of the thin brown husk, just as you wish. It makes no difference one way or another except that if it's off, you can see the first stages of development more easily. I usually leave it on.

Now you have a lovely avocado pit with arms radiating from its mid-circumference looking like a miniature botanical space craft. Take an ordinary drinking glass (normally four to six inches deep) and sit the pit on the rim. Fill the glass with water (warm—always tepid-to-warm water for avocados, even when watering the full-grown plant) so that, as I've said, about half an inch of the pit's bottom is covered. From now on, keep the water at that level. I had a frantic letter from one beginner saying, "The water's evaporating! What shall I do!" All I could think of in the way of advice was, "Get another hobby."

Beginning to root in a glass of water

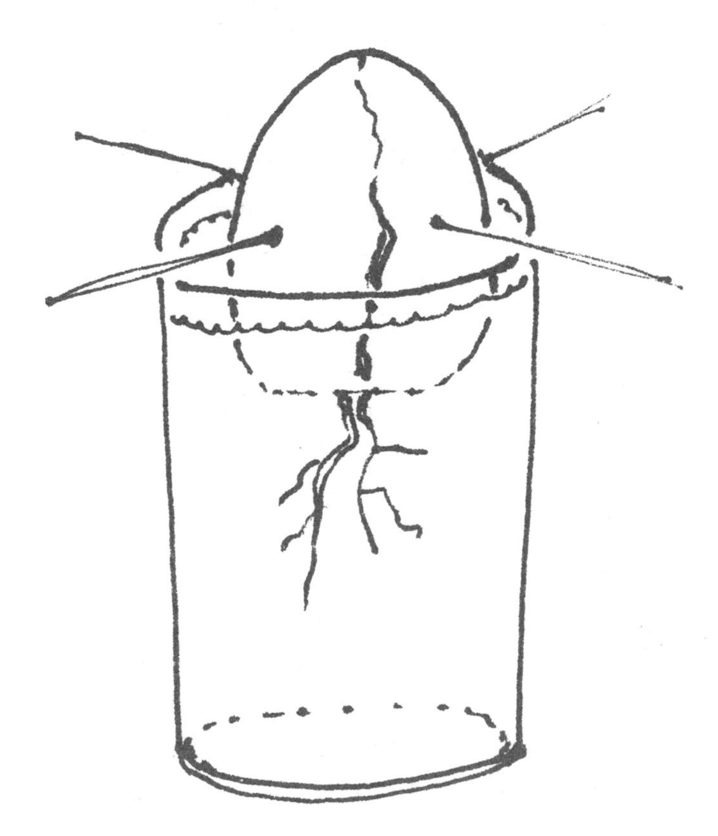

32

If you took the pit from a really ripe avocado, it was already splitting or even showing roots—in extreme cases, a sprout as well—and you were careful not to injure them while giving the pit its natal bath. More likely, since 99 per cent of avocados come onto the market half green and are eaten that way, the pit was a solid mass. Before long you will see it put down little white nubbins of roots from the navel into the water, and it will begin to split along the sides from the bottom up. How long it will take for this to happen (if it happens at all) depends on the whim of the avocado. Remember, you're dealing with a plant that has been hybridized and rehybridized so many times that there is no pure type any more—at least not in North American markets. Just be patient. But if the water clouds up and the pit begins to smell, you've got hold of a dud, a sterile pit, and it's rotting. Throw it out and begin over.

Whatever method you are using, you want a nice warm place (you're dealing with a native of the tropics) where the temperature doesn't change too much. Dark or light makes no difference until you get a sprout. After all, it's dark inside the pit where things begin happening. I have germinated pits by putting them in cellophane bags with a bit of sopping paper towel, closing them up tight, and keeping them on a bathroom shelf where they entertained me, in their quiet way, every morning as I shaved. I still keep them there in the early stages of the method I now like to use. My bathroom—my wife has her own—has a shelf for pots, earth, and other stuff and is known in our family as the potting shed.

Germinating in Earth

It is really most reasonable, although less of a show, to start your pit in earth in the first place. You can start with a pot of any size, but the rule is the larger the better, with an 8-inch pot (pots are measured by the diameter of the top) probably the minimum as a beginning. But as a personal eccentricity I like to begin smaller and repot frequently, perhaps just for the fun of it although I tell myself that

that way the roots make a series of compact balls, growing larger from pot to pot, that take full advantage of the available earth. I often begin with a pot as small as four-and-a-half inches filled with very moist earth to within half an inch of the top with the pit (Flat end down! Point end up!) sunk to about a quarter or at most a half its mass.

Once you've got your pit nicely settled you can just keep the soil very moist—you'll be amazed at how much the pit can drink—or do a variation on the cellophane bag trick, which is what I like to do. With a bag of an appropriate size you can just put the pot in and then tie up the end, or you can be fancier and lower the bag over the pot leaving a nice bubble of air at the top and closing the bag under the pot. You can string-tie or scotch-tape or rubber-band the bag close to the sides of the pot. If the earth is moist enough—it should be good and moist but not sopping—moisture will condense on the inside of the bag, and the pit will be in a heavenly sort of womb with 100 per cent humidity and, since you're keeping it out of the sun until it has a healthy sprout, steady temperature. Once the sprout appears, you're all set to put the plant in your best window. But in the meantime, we've left a pit all rooted and waiting in that water glass and must get back to it.

Artificial womb for germinating

—with this result about ten days later

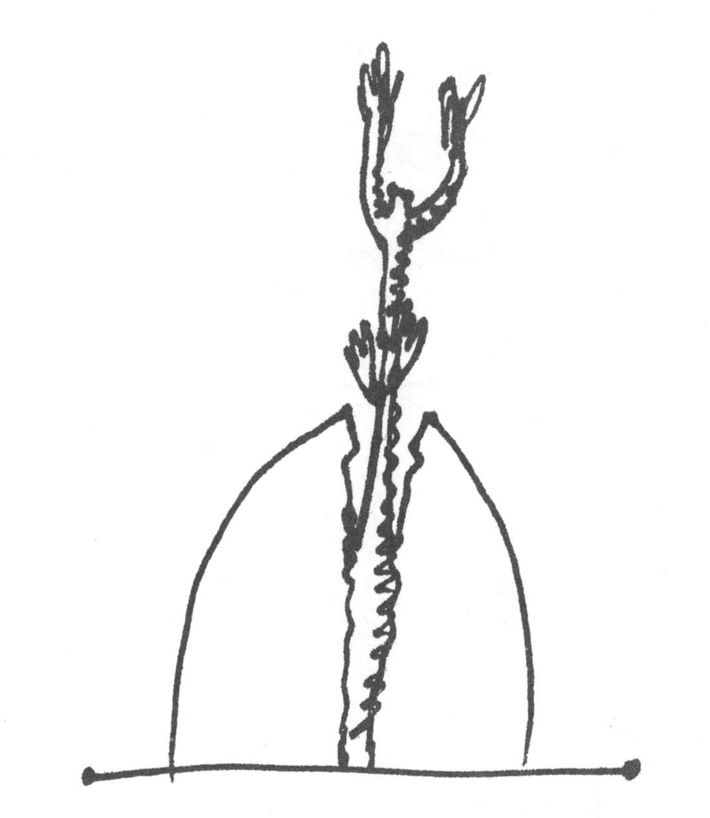

Potting and Repotting

By the time the roots of a water-germinated pit are reaching down toward the bottom of the glass and curling out toward the sides, you must put the plant into earth, although you can do it earlier if you wish. The trouble now is to get the roots into earth without breaking them. It's just a matter of being careful. The earth as it comes in bags from the store is fine-textured enough so that you can hold the rooted pit in position (you are going to bury it about half way) and sift the earth down around the roots until the pot is full.

There's another way. As soon as the roots begin to appear you can add a bit of earth daily to the water glass. It will sink to the bottom and grow solider as more is added until finally a ball of earth entirely replaces the water. A friend who lives in Italy tells me the Italians root cuttings this way in bottles, then break the bottles and put the ball of earth in the garden. When you have developed a solidish mass of earth around your avocado's roots, get rid of the glass and put the plant in a pot. It sounds reasonable, if you want to play it that way. To tell you the truth, I haven't tried it.

As a heretic I never bother with the layer of pot shards that all gardening books say you must put into the bottom of a pot. I put one shard over the hole to keep the earth from running out—loosely, so the water can run through—and that's that. When it comes time for repotting, if the earth won't come out easily, break the pot. It's the easy way, and not all that expensive until you get into great big pots. My mammoth avocado is now in a thirty-dollar pot, and it appears that I may have to break it to get the plant into a bigger one. But that's my problem. At the moment you're in a minor league.

As for the kind of pot: please use an ordinary clay one, the reddish kind. They are healthiest for the plant and, I think, the best looking. They age nicely, too. But avoid the clay saucers they will try to sell you to set them in. These too are porous—it's the porosity that makes the pot best for the plant—and they'll ruin anything they sit on indoors by collecting a puddle. You do need something to set the pot in. The simpler the better is my taste, and I have a stack of a dozen Pyrex baking dishes, ten inches in diameter and an inch and a half deep, always on hand to accommodate the shifting population of my window-sill plants.

The bigger the pot, the happier your plant will be in it, but there is the practical consideration of the amount of space you have, and the aesthetic one of the look of the plant in proportion to its pot. There seems to be a nice, automatic harmony that helps you determine when an avocado needs repotting. If it looks lost in its pot, the pot's more than roomy. As soon as it begins to look overgrown—so that you wonder how it can feed on so little earth, or looks as if it might turn over in a wind—it's probably time to repot.

A Few Basic Points at This Point

Now that your avocado is germinated, you've got to wait a few days or weeks to begin the pruning and shaping that may go on for years, so let's take advantage of the breather to establish a few fundamentals:

POTS: We have just considered this problem. Add, that for big plants you may want to consider wooden tubs. These are usually cedar and brass-bound. I happen not to like their looks, but that's personal. They're friendly to the plants.

EARTH: Avocados will grow in anything, I suspect, but they really like rich soil best. (Not all plants do.) Usually, I use potting soil from any garden supply house, although one florist sold me a bargain bushel of unpackaged earth just called "loam" that was a lot cheaper and worked fine. If you live in the country, just go out and dig up something nice and black.

WATER: Avocados like a lot. This is good, because why do you have house plants if not for the fun of caring for them? I water all my plants daily (warm for the avocados, quite cool for the cyclamens) and in the case of the avocados don't always observe the horiticultural warning that plants mustn't be allowed to stand in water ever. A plant with enough leaves (it's through the leaves that they lose water into the air—and, incidentally, my avocados make my apartment just about the best-humidified in New York City) will use up whatever runs out of the pot into the saucer after a heavy watering. I sometimes have to water my avocados twice a day. I've also put them in big pans holding enough water for a couple of days' ration when I have had to be away briefly, and they haven't drowned. Avocados don't like to dry out, although in an emergency they can survive a brief drought.

FERTILIZER: You can begin fertilizing even during the germinating period. Ever since a plant store manager told me that the chemical fertilizers are to natural fertilizers what a shot in the arm is to a good meal, I have stuck to his recommendation of fish emulsion, used according to directions on the bottle. But I know people who fertilize with what my adviser would regard as the equivalent of heroin, and their plants thrive. So take your choice. (Another thing is that overfertilizing with the chemical types can be fatal to a plant, while an overdose of natural fertilizer can be overcome.)

LIGHT: The more the better is the rule (for avocados). The only light-damaged avocado I've ever heard of is one that was subjected to an ultra-violet plant lamp twenty-four hours a day. The leaves began to yellow and drop off in exactly the way they would have if there hadn't been *enough* light. When the plant was given a resting period every night, it recovered nicely.

I'm leery of subjecting the newborn avocado shoot to strong sunlight on a window sill for the first week or so because a bright sun coming through a glass pane intensifies the heat. But even a baby avocado can soon take all the light you can give it. I've not much experience with artificial light for large plants, since I haven't room for equipment such as light stands. But as long as a plant is small enough to be given extra light by an ordinary lamp, I've had good results with ordinary electric bulbs either clear or frosted. I happen to have a couple of lamps built to accommodate small sculptures on their bases, and sometimes remove the shades and substitute a very young potted plant for the sculpture, and the plant thrives upon a few hours extra

An excellent type of lamp (ordinary 100-watt bulbs) for artificial light—but whatever you have will do

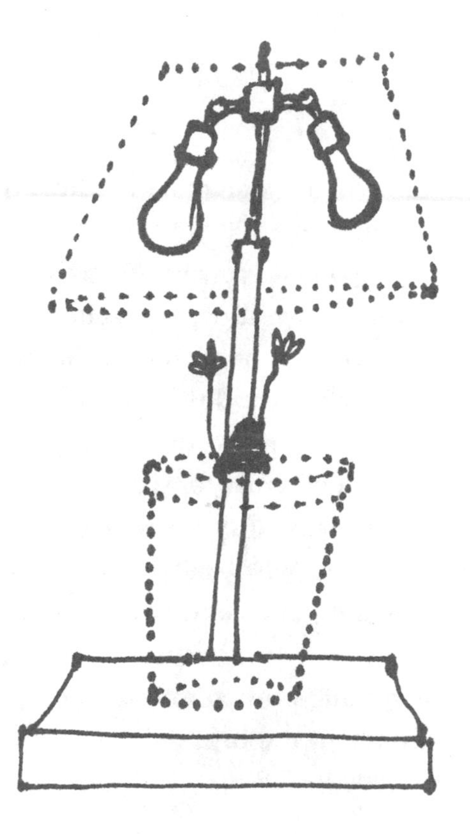

light a day. In the diagram on the previous page, the bulbs are 100-watt.

Depending on where you are growing your plant, you just have to rig up your own light supplements if you need them. I'm lucky to have very sunny windows. Whatever your light source, you have to turn your plant now and then or it will grow lopsided. I have some fast-growing plants that I turn systematically, 90 degrees clockwise daily, when watering them. They grow nice and straight as a result. But you can just turn a plant according to which side you want to encourage, giving it the advantage of the best light whenever you think it needs it.

Phase Two: Cutting Back and All That

Left to itself, an indoor avocado has just one ambition. It wants to beat the record for hitting the ceiling. It may throw out a skinny branch or two on the way up as an inadequate concession to looks, but essentially it will be a long underfed-looking stem crowned by a cluster of inadequate-looking leaves. You've got to chop it off sooner or later, not so much to keep it within bounds heightwise as to force it to branch out and make something of itself.

The best-established school of thought insists that the sprout must be cut down by half when it is five, six, or seven inches high. This seems to be a dependable rule, forcing a stronger stalk to emerge from the amputated

All it wants is to hit the ceiling

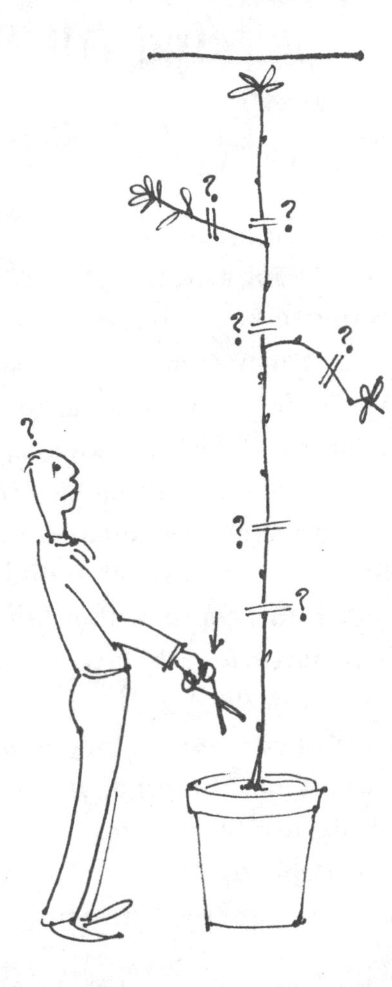

one. But I have cropped the sprouts much later and much earlier. At present I have a plant that was cropped about a quarter of an inch *below* the tip of the pit (reaching in with a pair of scissors to do it) when the top of the sprout was only about half an inch above the pit. This is very young, but the result so far is a triple renewal of the stalk. Three stalks of equal strength have emerged from the former single stalk and are a couple of inches high. Ask me about this one in a few years.

Branches as well as the main stem (or stalk or trunk, whatever you like to call it) can be clipped in the hope that foliage will crop out below the cut. Generally speaking, cutting a strong branch encourages the weaker ones, which take advantage of the big boy's temporary state of shock. The rule is that cutting the plant back will thicken and strengthen what is left of it, but you can't absolutely depend—in fact you can't depend at all—on an avocado's branching out exactly where you want it to. The juncture of each leaf with its branch is potentially the point of growth of another branch, and some of those little warty-looking spots on branches are also nodes (points of growth). But the avocado picks and chooses without much regard as to how you've tried to direct it. To make up for this, there is a fascination in watching it decide what shape it wants to take.

I have one avocado with a trunk about an

inch in diameter—I think it's about five years old—that I have put through a series of experimental traumas and recently cut down to within five inches of the earth. It immediately put out at least two dozen little green buds, and I thought for a while it was going to act like a privet, but most of the buds dried up. Three shot up fast and are still growing while a few others are marking time.

All in all, the best advice I can give about cutting back is improvise and don't be afraid of the scissors. Cutting back a skinny plant almost always results in improved appearance or, if not that, at least an interesting eccentricity. And after you've cut back, don't panic when the plant sulks. Cutting back slows things down, but the plant will soon gather its forces together again and go zooming along until you have to get the scissors out once more. One nice thing about avocados is they never harbor a grudge.

Sprout may be topped anywhere from 2 to 3-4 inches above pit. Here, about 2 inches

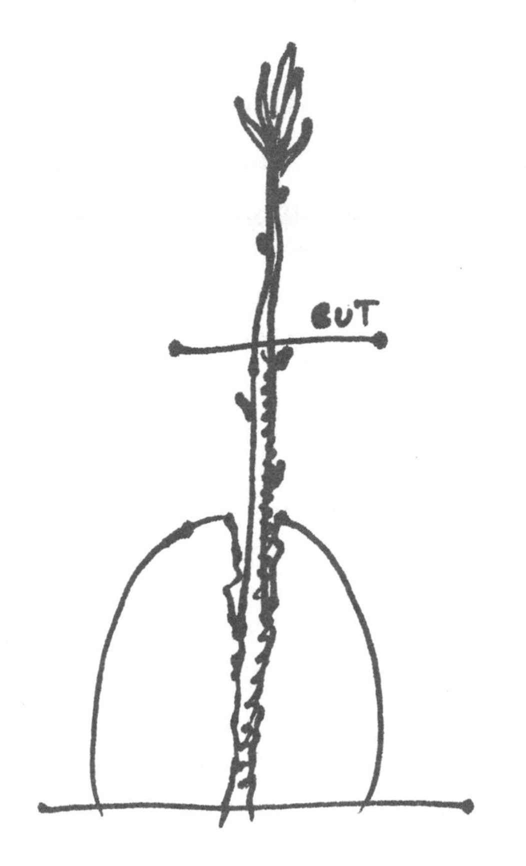

Fancy Stuff

The pliability of young avocado branches, upon which I capitalized for practical reasons in creating the Weeping Avocado effect already described, has tempted me to induce other deformities for purely aesthetic ends. Let's face it: The avocado will never rival the yew or boxwood as a topiary plant. But the branches can be trained, or forced, into skeletons of ball-like shapes that the leaves fill out, and probably into other shapes by anyone with enough patience.

My patience gives out regularly, but the impulse to experiment is reborn just as regularly so that I usually have an avocado under treatment. The one mentioned just above, with the three branches that shot up from "at

Same avocado once again as now in training-three new branches being trained in whorls, leaves omitted for clarity

least two dozen little green buds," is the current victim. As shown in the illustrations, I am curving the branches into three horizontal arcs that, just now, make a nice whorl. I hope they will continue to grow and become circles spiraling upward. What I am after is a solid column of foliage. The branches come from the trunk close enough to the earth so that at present they can be held down into position by large-size paper clips opened up to leave a hook at one end that goes over the branch at the top of a long spike that goes into the earth. For later problems, later solutions. (I have forgotten to mention that after the branch has been held into position by one device or another for a while, it accepts that

Avocado subjected to Japanese-torture type deformation

Same avocado following defoliation

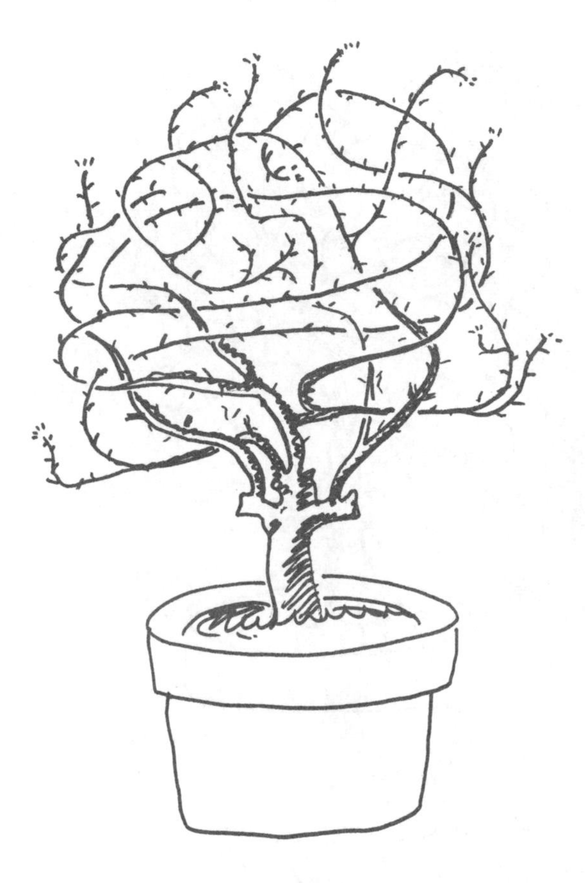

position as its natural one, and the restraints can be removed.)

The same plant was the subject of another experiment by which I managed to shape it into an elongated globe by intertwining the branches. But it was not really a happy plant. It took on a ragged look, and I tried defoliating it—picking off every leaf (and cutting off the tip of every branch) to encourage small branches to grow from the nodes. It worked more or less, but not enough. I cut the plant off as described, and began again. The drawings here are from photographs I took at the time—or times.

I just mention this fancy stuff in passing. It is fun, but sounds like something the Japanese would be better at.

Hints on Health and Hygiene

I have never had an avocado develop a blight, fungus, or similar malady. If one did, I wouldn't try to cure it with medicine but would just lop off all the affected parts and hope the new growth would be clean. Usually new growth inherits the malady, I have discovered from other plants. I truly think the only remedy for fungus and blight is not only to throw the plant out but to be sure it gets cremated to prevent infection of other plants.

As for insects, I've said all I have to say about them in the first part of this monograph. I don't worry about the various little creatures—springtails and such—that breed in the earth. I've never had any that hurt anything and their presence doesn't offend me.

They mind their own business, I mind mine.

So it comes to the leaves. Good color is the surest proof of a plant's good health, and bad color is the surest but, alas, least specific warning that something is wrong. Avocado leaves come out in a rosette of light green with a bronzish cast that is perfectly beautiful and quickly deepen to a pleasant, soft, uniform darker green. When a mature leaf begins to go yellow, it's beyond saving. A yellowing leaf never makes a comeback, so pick it off right away and get it over with.

My big avocado drops leaves the year round, but in various tempos—sometimes hardly any, sometimes more than I like. To a large extent this is seasonal. House plants by some miracle know what is going on outdoors even when they are in a constant temperature the year round from year to year of their lives. (Although I did manage once to fool a little maple tree and reverse its seasons by refrigerating it all summer and keeping it nice and warm all winter.)

When my avocados feel spring coming they burgeon at every point after growing no new leaves for months. Perhaps through some fault of mine many of these new leaves drop off before they mature. All my avocados seem to go in for excess and then come back into balance in a cycle of overachievement and withdrawal. But I always get enough new, healthy growth finally.

If you are really worried about too many yellow leaves, try repotting, drastic surgery, and probably less fertilizer rather than more. (Overfertilizing is always a temptation, especially in the months when a plant isn't growing much.) If the plant dies in spite of everything, then O.K.—it dies.

It is quite possible that yellowing during the summer months may come from too close proximity to an air conditioner. Gas leaks are another hazard for avocados as for all house plants. Then there is something I have never met up with but have secondhand experience of through friends—root strangulation. A potted plant can get its roots so intertwined that one, growing, can literally strangle another. This is a hazard in beginning plants in small pots, as I like to do, and is the reason why all repotting instructions always say "loosen the root ball," which I seldom bother to do. Just lucky, I guess.

Envoi and Gastronomical Notes

The avocados I've been talking about are the various predominantly green-skinned hybrids commonest in North American markets, although from time to time I have also planted pits from the smaller, blackish, blistered-looking variety. (Among other places, you meet these in central Africa, where I happen to have spent some time, and they are usually pretty stringy.) In Mexican markets the cooks shopping for avocados reject all hybrids contemptuously, calling them a name that sounds like "*wawa*" and has something of the ring of "gringo." What they are after is the pure Mexican variety, although I have never learned to identify it and for that matter I haven't noticed all that much difference in flavor between one avocado and another. But when it comes to guacamole, Mexicans are perfectionists.

So is my wife, and I suggested to her that she give me her guacamole recipe to wind up these notes with a fillip, but she said, "Ridiculous." My impression was that after everything she has gone through she didn't want any part of anything that might encourage other husbands to go in for indoor avocado culture, but after hemming and hawing for a while, which isn't like her, she finally admitted that the recipe isn't really hers, but comes from a friend of ours, Diana Kennedy, a towering figure in classical Mexican cookery who occasionally begs avocado leaves off me to toast and crush for flavoring in certain dishes. (They have a slight sort of licorice flavor, stronger in some varieties than others.)

Straight from Diana, the Mexican Escoffier, here is the classic guacamole recipe, beginning with the injunction, "Make it in a *molcajete* or pestle and mortar—please, no blenders."

2 medium avocados
½ medium onion, finely chopped
Leaves from 4 large sprigs of green coriander, finely chopped
1 or 2 fresh green chilies, finely chopped
¼ teaspoon salt, or to taste
1 large tomato

Mexican molcajete—(rough porous volcanic stone culinary mortar) "No blenders, please"

Cut open the avocados, remove the pits, and scoop out the flesh. Mash the flesh roughly. Grind 2 tablespoons onion, 2 coriander leaves, chilies, salt to form a paste, and mix well with the mashed avocado. Skin, seed, and chop the tomato well and add to the mixture. Add remaining 2 tablespoons of onion and 2 coriander leaves into the guacamole mixture. Serve immediately if possible.

Putting the pit into the guacamole while it stands helps to preserve the color a little, but if it simply has to stand, then cover tightly with plastic to exclude the air. *Buen provecho.*

Still in a confessional mood, my wife admitted that in a pinch or a hurry (meaning, no fresh green chiles and no green coriander) she gives the molcajete a scant sprinkling of salt and rubs it with garlic, puts the avocado in with a dash of lemon juice, mashes it not too smoothly with finely chopped onion, and uses chili powder out of a jar for seasoning—sometimes adding bits of tomato. More salt? Depends. Taste first. And the thinnest possible film of mayonnaise if it has to wait before it is eaten. Craig Claiborne may recognize the source of this variation, but my wife says he won't mind.

Swell, but don't let this lead you to think you are going to get any avocados off your indoor tree. Leaves, yes—fruit, no. If you do get fruit, call the nearest botanical garden with the news because you'll be the first. But you're going to have to do something with the avocados you buy for the pits, so I would like to say that in addition to guacamole, avocado makes a fine breakfast dish.

I'm not really a fan of the standard salad recipes, especially the combination of avocado slices (invariably not ripe enough) and grapefruit sections with French dressing. But many a time I've enjoyed breakfasting on an avocado served right in its half shell—creamy ripe and velvety soft—with no garnish except salt and coarsely ground pepper. After the sun is over the yardarm, add mustard sauce.

I also like half an avocado smothered in a chilled sauce of chopped cucumber and yoghurt. I invented this myself, and to tell you the plain truth, it's just wonderful.

Let me know if you discover anything extraordinary about avocados. You're likely to.

—And a Summary

Starting: Standard procedure is to suspend the avocado pit on the rim of a glass of water so that it is immersed for about half an inch. Fun to watch it put out roots, but you can also just plant the pit directly in earth, sinking it to about half its mass and keeping the earth wet.

Pots: Ordinary clay ones are always best, but you need waterproof saucers under them—preferably glass. Eight-inch pots are good for starters.

Earth: Buy potting earth in bags at the garden supply store.

Watering: Water, water, water, with water tepid to warm.

Fertilizers: Any commercial kind will do, but

fish emulsion has an edge on chemical types, which more easily "burn" the plant if too much is used.

REPOTTING: When the plant looks too big for its pot, it probably is. If plant and earth won't lift out of the pot when you grasp the trunk and pull, break the pot. Breaking is always safest, for that matter.

LIGHT: All the sunlight you can supply will be relished. If you use artificial light turn it off a few hours during the night.

PRUNING: Cutting back will always encourage new growth and encourage more leaves on existing growth. Cut back with this in mind and just hope the avocado feels like putting out the new growth where you want it.

Shaping: A young branch is pliable and can be bent into any position you like. After being held in place by strings or other devices for a few weeks or a couple of months it will have stiffened into the position and restraints can be removed.

Disease: Difficult question. Best to destroy badly diseased plants. Yellowing leaves are symptoms of various kinds of ill health when they are not just a seasonal matter. Pruning, repotting, a warmer temperature (and always water) are some obvious means to help a sick plant, but sometimes you just can't tell, and it dies.

Over-all remedy: Plant another.